ICME-13 Topical Surveys

Series editor

Gabriele Kaiser, Faculty of Education, University of Hamburg, Hamburg, Germany

More information about this series at http://www.springer.com/series/14352

Carmen Batanero · Egan J. Chernoff
Joachim Engel · Hollylynne S. Lee
Ernesto Sánchez

Research on Teaching and Learning Probability

 Springer Open

Carmen Batanero
Facultad de Ciencias de la Educación
University of Granada
Granada
Spain

Egan J. Chernoff
College of Education
University of Saskatchewan
Saskatoon, SK
Canada

Joachim Engel
Department of Mathematics
and Computer Science
Ludwigsburg University of Education
Ludwigsburg
Germany

Hollylynne S. Lee
Department of Science, Technology,
 Engineering, and Mathematics
 Education
North Carolina State University
Raleigh, NC
USA

Ernesto Sánchez
Departamento de Matemática Educativa
CINVESTAV-IPN
Mexico, Distrito Federal
Mexico

ISSN 2366-5947   ISSN 2366-5955   (electronic)
ICME-13 Topical Surveys
ISBN 978-3-319-31624-6   ISBN 978-3-319-31625-3   (eBook)
DOI 10.1007/978-3-319-31625-3

Library of Congress Control Number: 2016937348

Printed on acid-free paper

This Springer imprint is published by Springer Nature
The registered company is Springer International Publishing AG Switzerland

Main Topics You Can Find in This "ICME-13 Topical Survey"

In this topical survey, we present a summary of some of the most important and recent work in probability education. In this survey, we briefly summarise the following:

- Analysis of the nature of chance and probability;
- Main components of probabilistic knowledge and reasoning;
- Analysis of probability in the school curricula;
- Intuitions and learning difficulties in probability;
- Technology and educational resources in teaching and learning probability;
- Education of teachers for teaching probability.

Contents

Research on Teaching and Learning Probability

1 Introduction

To adequately function in society, citizens need to overcome their deterministic thinking and accept the existence of fundamental chance in nature. At the same time, they need to acquire strategies and ways of reasoning that help them in making adequate decisions in everyday and professional situations where chance is present.

This need for probability literacy has been recognized by educational authorities in many countries by including probability in the curricula at different educational levels and in the education of teachers. However, including a topic in the curriculum does not automatically assure its correct teaching and learning; the specific characteristics of probability, such as a multifaceted view of probability or the lack of reversibility of random experiments, are not usually found in other areas and will create special challenges for teachers and students.

Research in probability education tries to respond to the above challenges and it is now well established, as shown by the Teaching and Learning of Probability Topic Study Group at the 13th International Congress of Mathematics Education (ICME). Probability education research is also visible in the many papers on this topic presented at conferences such as the European Mathematics Education Conference (CERME), the International Conference on Teaching Statistics (ICOTS), as well as in regional or national conferences such as the Latin-America Mathematics Education Conference (RELME).

Furthermore, several books and major handbook chapters listed in the Further Readings section suggest the relevance of this field and the need to (re)formulate a research agenda in this area for the coming years. In this survey on the state of the art, we summarise existing research in probability education before pointing to some ideas and questions that may help in framing a future research agenda.

C. Batanero, E. Chernoff, J. Engel, H. Lee, and E. Sánchez

© The Author(s) 2016
C. Batanero et al., *Research on Teaching and Learning Probability*,
ICME-13 Topical Surveys, DOI 10.1007/978-3-319-31625-3_1

2 Survey on the State of the Art

Research in probability education has a fairly long history and includes theoretical analyses and empirical research on a variety of topics and from different perspectives, as described in the next sections. As we reviewed the existing literature on probability education, several major themes came to the fore. These themes have been used to organize a brief review of our current understanding of probability education that informs the discussions for the 2016 ICME topic study group.

2.1 The Nature of Chance and Probability

Research in any area of mathematics education should be supported by an epistemological reflection about the objects that are being investigated. This reflection is especially relevant when focusing on probability, where different approaches to the concept that influence both the practice of stochastics and the school curricula are still being debated in the scientific community.

According to Hacking (1975), probability was conceived from two main, albeit different, perspectives since its emergence. A *statistical side* of probability is related to the need to find the objective mathematical rules that govern random processes; probability values are assigned through data collected from surveys and experiments. Complementary to this vision, an *epistemic side* views probability as a personal degree of belief, which depends on information available to the person assigning a probability. From these two main perspectives, which are reflected in the works of the main authors who have contributed to the progress of probability, different views of probability have been sustained through history (Batanero 2015; Batanero and Díaz 2007; Batanero et al. 2005a, b; Borovcnik and Kapadia 2014a; Chernoff and Russell 2012). Currently, the main primary interpretations are intuitive, classical, frequentist, subjective, logical, propensity, and axiomatic. Each of these views entails some philosophical issues and is more suited to model particular real-world phenomena or to be taken into account in curricula for specific students.

In the next sections we briefly summarise the main features of the aforementioned views of probabilities, part of which have been introduced in school curricula.

2.1.1 Intuitive View of Probability

The theory of probability is, in essence, a formal encapsulation of intuitive views of chance that lead to the fundamental idea of assigning numbers to uncertain events. Intuitive ideas about chance emerged very early in history in many different cultures and were linked to problems related to setting fair betting in games of chance (Batanero and Díaz 2007; Bennet 1999). According to David (1962), cubic dice were abundant in primitive cultures (e.g., the Egyptian, Chinese, Greek and Roman civilizations), which used games of chance in an attempt to predict or control fate in decision-making or religious ceremonies. Interestingly, the development of the theory of probability is much more recent with, according to David (1962), no clear reasons to explain this delay.

Intuitive ideas about chance and probability also appear in young children who use qualitative expressions (such as terms "probable" or "unlikely") to express their degrees of belief in the occurrence of random events. These intuitive ideas can be used by a teacher to help children develop a more mature understanding and use probability as a tool to compare likelihood of different events in a world filled with uncertainty.

2.1.2 Classical Meaning

The earlier theoretical progress in probability theory was linked to games of chance such as throwing dice. For example, in his correspondence with Fermat, Pascal (1654/1963) solved the problem of estimating the fair amount to be given to each player if the game is interrupted by "force majeure" by proportionally dividing the stakes among each player's chances. In another example, Cardano (1663/1961) advised players to consider the number of total possibilities and the number of ways favourable results can occur and to compare the two numbers in order to make a fair bet.

It is not surprising that the initial formalization of this concept was based on an assumption that all possible elementary events were equiprobable, since this hypothesis is reasonable in many chance games. In the classical definition of probability, given by Abraham de Moivre in 1718 in *Doctrine of Chances* and later refined by Laplace in 1814 in his *Philosophical Essay on Probability*, probability is simply a fraction of the number of favourable cases to a particular event divided by the number of all cases possible. This definition has been widely criticised since its publication, since the assumption of equiprobability of outcomes is subjective and it impedes the application of probability to a broad variety of natural phenomena where this assumption may not be valid.

2.1.3 Frequentist Meaning

The convergence of relative frequencies for the same event to a constant value after a large number of independent identical trials of a random experiment has been observed by many authors. In trying to extend the scope of probability to life-expectancy and insurance problems, Bernoulli (1713/1987) proved a first version of the *Law of Large Numbers*. According to this theorem, the relative frequency h_n for a given event in a large number of trials should be close to the theoretical probability p of that event and tend to become closer as more trials are performed.[1] Given that stabilised frequencies are observable, this theorem was also considered as a proof of the objective character of probability (Fine 1971).

In this frequentist approach, sustained later by von Mises (1928/1952) and Renyi (1966/1992), probability is defined as the hypothetical number towards which the relative frequency tends when a random experiment is repeated infinitely many times.

Since such an empirical tendency is visible in many natural phenomena, this particular definition of probability extended the range of applications enormously.

A practical drawback of this frequentist view is that we only obtain an estimation of probability that varies from one series of repetitions of experiments (called samples) to another. Moreover, this approach is not appropriate when it is not possible to repeat an experiment under exactly the same conditions (Batanero et al. 2005a, b). Consequently, it is important to make clear to students the difference between a theoretical model of probability and the frequency data from reality used to create a model of probability. Sometimes this difference is not made explicit in the classroom and may confuse students who need to use abstract knowledge about probability to solve concrete problems from real life.

2.1.4 Propensity Meaning

Popper (1959) introduced the idea of *propensity* as a measure of the tendency of a random system to behave in a certain way and as a physical disposition to produce an outcome of a certain kind. In the same sense, Peirce (1910/1932) proposed a concept of probability according to which a die, for example, possesses expected dispositions for its various possible outcomes; these propensities are directly related to the long-run trends and indirectly to singular events.

In the long run, propensities are tendencies to produce relative frequencies with particular values, but the propensities are not the probability values themselves (Gillies 2000). For example, a cube-shaped die has an extremely strong tendency (i.e., propensity) to produce a 5 when rolled with long-run relative frequency of 1/6. The probability value 1/6 is small, so it does not measure this strong tendency. In single-case theory (e.g., Mellor 1971) the propensities are identical to the probability values and are considered as probabilistic causal tendencies to produce a particular result on a specific occasion.

Again this propensity interpretation of probability is controversial. In the long-run interpretation, propensity is not expressed in terms of other empirically verifiable quantities, and we then have no method of empirically finding the value of a propensity. With regards to the single-case interpretation, it is difficult to assign an objective probability for single events (Gillies 2000). It is also unclear whether single-case propensity theories obey the probability calculus or not.

---

[1] Given $\varepsilon > 0$, $\alpha > 0$ arbitrarily small, the theorem establishes that for $n > \frac{pq}{\varepsilon\alpha^2}$, with q = 1−p $P(|h_n - p| < \varepsilon) \geq 1 - \alpha$.

2.1.5 Logical Meaning

Researchers such as Keynes (1921) and Carnap (1950) developed the logical theories of probability, which retain the classical idea that probabilities can be determined a priori by an examination of the space of possibilities; however, the possibilities may be assigned unequal weights. In this view, probability is a degree of implication that measures the support provided by some evidence E to a given hypothesis H. Between certainty (1) and impossibility (0), all other degrees of probability are possible. This view amplifies deductive logic, since implication and incompatibility can be considered as extreme cases of probability.

Carnap (1950) constructed a formal language and defined probability as a rational degree of confirmation. The degree of confirmation of one hypothesis H, given some evidence E, is a conditional probability and depends entirely on the logical and semantic properties of H and E and the relations between them. Therefore, probability is only defined for the particular formal language in which these relations are made explicit.

Another problem in this approach is that there are many possible confirmation functions, depending on the possible choices of initial measures and on the language in which the hypothesis is stated. A further problem is selecting the adequate evidence E in an objective way, since the amount of evidence might vary from one person to another (Batanero and Díaz 2007).

2.1.6 Subjective Meaning

In the previous approaches presented, probability is an "objective" value that we assign to each event. However, Bayes' theorem, published in 1763, proved that the probability for an event can be revised in light of new available data. A simple version this theorem establishes that, when the "prior" probabilities $P(A_i)$ and the likelihood $P(B|A_i)$ to obtain B for each A_i are known for a number of incompatible events A_i such that $\bigcup_{i=1}^{n} A_i = E$, then it holds:

$$P(A_i|B) = \frac{P(B|A_i) \cdot P(A_i)}{\sum_{j=1}^{n} P(B|A_j) \cdot P(A_j)}$$

Using Bayes' theorem, an initial (prior) probability can be transformed into a posterior probability using new data and probability loses its objective character. Following this interpretation, some mathematicians (e.g., Keynes, Ramsey and de Finetti) considered probability as a personal degree of belief that depends on a person's knowledge or experience. However, the status of the prior distribution in this approach was criticised as subjective, even if the impact of the prior diminishes by objective data, and de Finetti proposed a system of axioms to justify this view in 1937.

In this subjectivist viewpoint, the repetition of the same situation is no longer necessary to give a sense to probability, and for this reason the applications of probability entered new fields such as politics and economy, where it is difficult to assure replications of experiments. Today the Bayesian approach to inference, which is based in this approach, is quickly gaining further traction in numerous fields.

2.1.7 Axiomatic Theory

Despite the strong philosophical discussion on the foundations, the applications of probability to all sciences and sectors of human activity expanded very quickly. Throughout the 20th century, different mathematicians tried to formalise the mathematical theory of probability. Following Borel's work on set and measure theory, Kolmogorov (1933/1950), who corroborated the frequentist view, derived an axiomatic theory.

The set S of all possible outcomes of a random experiment is called the sample space of the experiment.[2] In order to define probability a set algebra A containing subsets of the sample space and which is closed under numerable union and complement operations is considered.[3] The complement of an event $\bar{A}$ is made of all the outcomes that do not take part in A. The event $S = A \cup \bar{A}$ always happens and is called a certain event.

Probability is any function defined from A in the interval of real numbers $[0,1]$ that fulfils the following three axioms, from which many probability properties and theorems can be deduced:

1. $0 \leq P(A) \leq 1$, for every $A \in A$;
2. $P(S) = 1$;
3. (a) For a finite sample space S and incompatible or disjoint events A and B, i .e., $A \cap B = \emptyset$, it holds that $P(A \cup B) = P(A) + P(B)$.
   (b) For an infinite sample space S and a countable collection of pairwise disjoint sets $A_i, i = 1,2,\ldots$ it holds, $P\left(\bigcup_{i=1}^{\infty} A_i\right) = \sum_{i=1}^{\infty} P(A_i)$.

This axiomatic theory was accepted by the different probability schools because, with some compromise, the mathematics of probability (classical, frequentist or subjective) may be encoded by Kolmogorov's theory. However, the interpretation of what is a probability would differ according to the perspective one adheres to; the discussion about the meanings of probability is still very much alive in different approaches to statistics. This link between probability and philosophy may also explain people's intuitions that often conflict with the mathematical rules of probability (Borovcnik et al. 1991).

---

[2]The sample space may be finite, countably infinite or uncountably infinite. The sample space is countably infinite when it can be put in correspondence with the set N of natural numbers.

[3]For finite or countably infinite sample spaces S, A includes all subsets of S. For finite sample spaces A is a Boolean Algebra; in case of a countably infinite sample space, A is a σ-algebra, that is, it is closed under countable union and intersection operations and complement building. For uncountable (continuous) infinite sample spaces, the set algebra considered to assign probability does not include all possible subsets of S; A is restricted instead to a system of events of S, closed under countable union and intersection operations and forming complements (Chung 2001).

2.1.8 Summary of Different Views

Our exposition suggests that the different views of probability described involve specific differences, not only in the definition of probability itself, but also in the related concepts, properties, and procedures that have emerged to solve various problems related to each view. We summarise some of these differences in Table 1, partially adapted from Batanero and Díaz (2007).

Table 1 Elements characterizing the different views of probability (adapted from Batanero and Díaz 2007, p. 117)

Views of probability	Procedures	Properties	Some related concepts
Classical	• Combinatorics • Proportions • A priori analysis of the experiment structure	• Proportion of favourable to all possible cases • Equiprobability of elementary events	• Expectation • Fairness
Frequentist	• A posteriori collection of statistical data • Statistical analysis of data • Curve fitting	• "Limit" of relative frequencies in the long run • Objective; based on empirical facts • Repeatable experiment	• Relative frequency • Data distribution • Convergence • Independence of trials
Propensity	• A priori analysis of the experimental set up	• Physical disposition or tendency • Applicable to single cases • Related to the experimental conditions	• Propensity • Probabilistic causal tendency
Logical	• A priori analysis of the space of possibilities • Propositional logic • Inductive logic	• Objective degree of belief • Revisable with experience • Relationships between two statements, generalises implication	• Evidence • Hypothesis • Degree of implication
Subjective	• Bayes' theorem • Conditional probability	• Subjective character • Revisable with experience	• Likelihood • Exchangeability • A priori probability (or distribution) • A posteriori probability (or distribution)
Axiomatic	• Set theory • Set algebra	• Measurable function	• Sample space • Certain event • Algebra of events • Measure

2.1.9 Different Views of Probability in School Curricula

The above debates were, and are, reflected in school curricula, although not all the approaches to probability received the same interest. Before 1970, the classical view of probability based on combinatorial calculus dominated the secondary school curriculum in countries such as France (Henry 2010). Since this view relies strongly on combinatorial reasoning, the study of probability, beyond very simple problems, was difficult for students.

The axiomatic approach was also dominant in the modern mathematics era because probability was used as a relevant example of the power of set theory. However, in both the classical and axiomatic approaches, multiple applications of probability to different sciences were hidden to students. Consequently, probability was considered by many secondary school teachers as a subsidiary part of mathematics, dealing only with chance games, and there was a tendency to "reduce" the teaching of probability (Batanero 2015).

Today, with the increasing interest in statistics and technology developments, the frequentist approach is receiving preferential treatment. An experimental introduction of probability as a limit of relative frequencies is suggested in many curricula and standards documents (e.g., the Common Core State Standards in Mathematics [CCSSI] 2010; the Ministerio de Educación, Cultura y Deporte [MECD] 2014; and the National Council of Teachers of Mathematics [NCTM] 2000), and probability is presented as a theoretical tool used to approach problems that arise from statistical experiences. At the primary school level, an intuitive view, where children start from their intuitive ideas related to chance and probability, is also favoured. The axiomatic approach is not used at the school level, being too formal and adequate only for those who follow studies of pure mathematics at the post-secondary level. More details of probability contents in the school curricula will be discussed in Sect. 2.3.

2.2 Probabilistic Knowledge and Reasoning

The recent emphasis on the frequentist view and on informal approaches in the teaching of inference may lead to a temptation to reduce teaching probability to the teaching of simulations—with little reflection on probability rules. However, as described by Gal (2005), probability knowledge and reasoning is needed in everyday and professional settings for all citizens in decision-making situations (e.g., stock market, medical diagnosis, voting, and many others), as well as to understand sampling and inference, even in informal approaches. Moreover, when considering the training of scientists or professionals (e.g., engineers, doctors) at university level, a more complex knowledge of probability is required. Consequently, designing educational programmes that help develop probability knowledge and reasoning for a variety of students requires the description of its different components.

While there is an intense discussion on the nature of statistical thinking and how it differs from statistical reasoning and statistical literacy (e.g., Ben-Zvi and Garfield 2004), the discussion of core components of probabilistic reasoning is still a research concern. Below we describe some points to advance future research on this topic.

2.2.1 What Is Probabilistic Reasoning?

Probability constitutes a distinct approach to thinking and reasoning about real-life phenomena. Probabilistic reasoning is a mode of reasoning that refers to judgments and decision-making under uncertainty and is relevant to real life, for example, when evaluating risks (Falk and Konold 1992). It is thinking in scenarios that allow for the exploration and evaluation of different possible outcomes in situations of uncertainty. Thus, probabilistic reasoning includes the ability to:

- Identify random events in nature, technology, and society;
- Analyse conditions of such events and derive appropriate modelling assumptions;
- Construct mathematical models for stochastic situations and explore various scenarios and outcomes from these models; and
- Apply mathematical methods and procedures of probability and statistics.

An important step in any application of probability to real-world phenomena is modelling random situations (Chaput et al. 2011). Probability models, such as the binomial or normal distribution, supply us with the means to structure reality: they constitute important tools to recognise and to solve problems. Probability-related knowledge relevant to understanding real-life situations includes concepts such as conditional probabilities, proportional reasoning, random variables, and expectation. It is also important to be able to critically assess the application of probabilistic models of real phenomena. Since today an increasing number of events are described in terms of risk, the underlying concepts and reasoning have to be learned in school, and the understanding of risk by children should also be investigated (Martignon 2014; Pange and Talbot 2003).

2.2.2 Paradoxes and Counterintuitive Results

Probabilistic reasoning is different from reasoning in classical two-valued logic, where a statement is either true or false. Probabilistic reasoning follows different rules than classical logic. A famous example, where the transitivity of preferences does not hold, is Efron's intransitive dice (Savage 1994), where the second person who selects a die to play always has an advantage in the game (no matter which die their opponent first chooses).

Furthermore, the field of probability is replete with intuitive challenges and paradoxes, while misconceptions and fallacies are abundant (Borovcnik and Kapadia 2014b). These counterintuitive results also appear in elementary probability, while in other areas of mathematics counterintuitive results only happen when working with advanced concepts (Batanero 2013; Borovcnik 2011). For example, it is counterintuitive that obtaining a run of four consecutive heads when tossing a fair coin does not affect the probability that the following coin flip will result in heads (i.e., the gambler's fallacy).

Probability utilises language and terminology that is demanding and is not always identical to the notation common in other areas of mathematics (e.g., the use of Greek or capital letters to denote random variables). Yet, probability provides an important thinking mode on its own, not just a precursor of inferential statistics. The important contribution of probability to solve real problems justifies its inclusion into school curriculum.

2.2.3 Causality and Conditioning

Another component of probabilistic reasoning is distinguishing between causality and conditioning. Although independence is mathematically reduced to the multiplicative rule, a didactical analysis of independence should include discussion of the relationships between stochastic and physical independence and of psychological issues related to causal explanation that people often relate to independence (Borovcnik 2012). While dependence in probability characterises a bi-directional relation, the two directions involved in conditional probabilities have a completely different connotation from a causal standpoint. For example, whereas the conditional probability of having some virus to having a positive result on a diagnostic test is causal, the backward direction of conditional probability from a positive diagnosis to actually having the virus is merely indicative. Alternatively stated, while the test is positive because of a disease, no disease is caused by a positive test result.

In many real-life situations the causal and probabilistic approach are intermingled. Often we observe phenomena that have a particular behaviour due to some causal impact factors plus some random perturbations. Then the challenge, often attacked with statistical methods, is to separate the causal from the random influence. A sound grasp of conditional probabilities is needed to understand all these situations, as well as for a foundation for understanding inferential statistics.

2.2.4 Random and Causal Variation

Another key element in probabilistic reasoning is discriminating random from causal variation. Variability is a key feature of any statistical data, and understanding of variation is a core element of statistical reasoning (Wild and Pfannkuch 1999). However, whereas variation of different samples from the same population or process (e.g., height of different students) may be attributed to random effects, the differences between samples from different populations (e.g., heights of boys and girls) are sometimes explained causally. Besides, the larger the size of the individual variation, the smaller the amount of variation that can be attributed to systematic causes.

A helpful metaphor in this regard is to separate the signal (the true causal difference) from the noise (the individual random variation) (Konold and Pollatsek 2002). Said authors characterise data analysis as the search for signals (causal

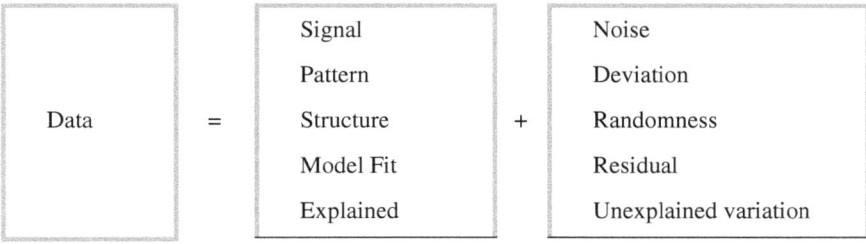

Fig. 1   Different versions of the structural equation

variations) in noisy processes (which include random variation). Borovcnik (2005) introduced the *structural equation,* which represents data as decomposed into a signal to be recovered and noise. Figure 1 displays five expressions of the signal-noise idea from different perspectives. The structural equation is a core idea of modelling statistical data and is a metaphor for our human response to deal with an overwhelming magnitude of relevant and irrelevant information contained in observed data. How to separate causal from random sources of variation is by no means unique. Probability hereby acquires more the character of a heuristic tool to analyse reality.

Random and causal sources of variation are complementary to each other, as they are considered in probability models used in statistical decision processes. Consider, for example, the problem of deciding whether expected values of two random variables differ. Several realisations of each of the two single variables will not be identical and most likely the empirical means will not be equal. Based on a sample of realisations of each random variable, we perform an analysis that leads to the classical two-sample statistical test. Statistical inference based on probabilistic reasoning provides methods and criteria to decide, with a margin of error, when the observed differences are due to random or causal variation.

It may be surprising, and from an epistemological point of view is far from obvious, that patterns of variation in careful measurements or in data of many individuals can be described by the same type of mathematics that is used to characterise the results of random experiments. Indeed, it is here where data and chance (i.e., statistics as the science of analysing data and probability as the study of random phenomena) come together to build the powerful foundation of statistical inference.

However, the above is not obvious for some students, who may reveal a prevailing inclination to attribute even small variation in observed phenomena to deterministic causes. As is the case in the following quote from a 17-year-old student: "I accept the idea of randomness when I ask for the sum of two dice, but what is random about the weight loss of a person following a particular diet plan?" (Engel et al. 2008). A perspective of losing weight as a noisy process may solve the problem for the student: sticking to a particular diet plan may have an influence on body weight over time, described by a (deterministic) function which, however, is affected by individual, unforeseen, and unpredictable random influences.

Wild and Pfannkuch (1999) state that people have a strong natural tendency to search for specific causes. This tendency leads people to search for causes even when an individual's data are quite within the bounds of expected random variation. Konold (1989) has accounted for this tendency in his *outcome approach*. This tendency is, in particular, visible in secondary school students, whose adherence to a mechanistic-deterministic view of the world is well documented and does not seem to fade with increasing years of schooling (Engel and Sedlmeier 2005).

2.2.5 Probabilistic Versus Statistical Reasoning

To conclude this section we remark that probabilistic reasoning is closely related to, and yet different from, statistical reasoning. Statistics can be portrayed as the science of learning from data (Moore 2010). At first glance it may be surprising to recognize that data (from Latin *datum*, the given) can be connected with randomness as the unforeseen. The outcome of a random experiment is uncertain. How is it possible to associate measurement readings collected in a concrete physical context with the rather metaphysical concept of randomness, which even cannot be defined in exact mathematical terms?

While probabilistic reasoning aims at structuring our thinking through models, statistical reasoning tries to make sense of observed data by searching for models that may explain the data. Probabilistic reasoning usually starts with models, investigates various scenarios and attempts to predict possible realizations of random variables based on these models. The initial points of statistical reasoning are data, and suitable models are fitted to these data as a means to gain insight into the data-producing process. These different approaches may be reconciled by paraphrasing Immanuel Kant's famous statement, "Theory without data is empty, data without theory is blind." Both statistical reasoning and probabilistic reasoning alone have their limitations and their merits. Their full power for advancing human knowledge comes to bear only in the synthesis acknowledging that they are two sides of the same coin.

2.3 Probability in School Curricula

The described need to understand random phenomena and to make adequate decisions when confronted with uncertainty has been recognised by many educational authorities. Consequently, the teaching of probability is included in curricula in many countries during primary or secondary education. An important area of research in probability education is the analysis of curricular guidelines and curricular materials, such as textbooks. Both topics are now commented on in turn.

2.3.1 Probability in Primary School

At the beginning of this century, the *Standards* of the National Council of Teachers of Mathematics (NCTM 2000) in the United States included the following recommendations related to understanding and applying basic concepts of probability for children in Grades 3–5:

- Describe events as likely or unlikely and discuss the degree of likelihood using such words as certain, equally likely, and impossible;
- Predict the probability of outcomes of simple experiments and test the predictions;
- Understand that the measure of the likelihood of an event can be represented by a number from 0 to 1.

Their recommendations have been reproduced in other curricular guidelines for Primary school. For example, in Spain (Ministerio de Educación y Ciencia [MEC] 2006), the language of chance and the difference between "certain," "impossible," and "possible" were introduced in Grades 1–2; in Grades 3–4 the suggestion was that children were encouraged to perform simple experiments and evaluate their results; and in Grades 5–6, children were expected to compare the likelihood of different events and make estimates for the probability of simple situations.

Today, some curricula include probability from the first or second levels of primary education (e.g., Australian Curriculum, Assessment and Reporting Authority [ACARA] 2010; MECD 2014, 2015; Ministerio de Educación Pública [MEP] 2012; Ministry of Education [ME] 2007), while in other curricular guidelines probability has been delayed to either Level 6 or to secondary education (e.g., CCSSI 2010; Secretaría de Educación Pública [SEP] 2011). In the case of Mexico, for example, probability was postponed to the middle school level on the argument that primary school teachers have many difficulties in understanding probability and therefore are not well prepared to teach the topic.

A possible explanation for the tendency to delay teaching probability is its diminished emphasis in some statistics education researchers' suggestions that statistical inference be taught with an informal approach. This change does not take into account, however, the relevance of educating probabilistic reasoning in young children, which was emphasised by Fischbein (1975), or the multiple connections between probability and other areas of mathematics as stated in the Guidelines for Assessment and Instruction in Statistics Education (GAISE) for pre-K-12 levels (Franklin et al. 2007, p. 8): "Probability is an important part of any mathematical education. It is a part of mathematics that enriches the subject as a whole by its interactions with other uses of mathematics."

2.3.2 Probability at the Middle and High School Levels

There has been a long tradition of teaching probability in middle and high school curricula where the topics taught include compound experiments and conditional probability. For example, the NCTM (2000) stated that students in Grades 6–8 should:

- Understand and use appropriate terminology to describe complementary and mutually exclusive events;
- Use proportionality and a basic understanding of probability to make and test conjectures about the results of experiments and simulations;
- Compute probabilities for simple compound events, using such methods as organised lists, tree diagrams, and area models.

In Grades 9–12 students should:

- Understand the concepts of sample space and probability distribution and construct sample spaces and distributions in simple cases;
- Use simulations to construct empirical probability distributions;
- Compute and interpret the expected value of random variables in simple cases;
- Understand the concepts of conditional probability and independent events;
- Understand how to compute the probability of a compound event.

Similar content was included, even reinforced, in other curricular guidelines, such as ACARA (2010), Kultusministerkonferenz (KMK) (2004, 2012), MEC (2007a, b), MECD (2015), and SEP (2011). For example, in Spain and South Australia the high school curriculum (MEC 2007b; MECD 2015; Senior Secondary Board of South Australia (SSBSA) 2002) for social science students includes the binomial and normal distributions and an introduction to inference (sampling distributions, hypothesis tests, and confidence intervals). In Mexico, there are different high school strands; in most of them a compulsory course in probability and statistics is included. In France, the main statistical content in the last year of high school (*terminale*, 17-year-olds) is statistical inference, e.g., confidence intervals, intuitive introduction to hypothesis testing (Raoult 2013). In the last level of high school, CCSSI (2010) also recommends that U.S. students use sample data and simulation models to estimate a population mean or proportion and develop a margin of error and that they use data from randomised experiments to compare two treatments and decide if differences between parameters are significant.

2.3.3 Fundamental Probabilistic Ideas

A key point in teaching probability is to reflect on the main content that should be included at different educational levels. Heitele (1975) suggested a list of fundamental probabilistic concepts that played a key role in the history of probability and are the basis for the modern theory of probability. At the same time, people frequently hold incorrect intuitions about their meaning or application in absence of instruction. This list includes the ideas of random experiment and sample space, the addition and multiplication rule, independence and conditional probability, random variables and distribution, combinations and permutations, convergence, sampling, and simulation. Below we briefly comment on some of these ideas, which were analysed by Batanero et al. (2005a, b):

- *Randomness.* Though randomness is a foundational concept in probability (random variation, random process, or experiment), it is a "fuzzy" concept, not always defined in textbooks. Research shows the coexistence of different interpretations as well as misconceptions held by students and suggests the need to reinforce understanding of randomness in students (Batanero 2015).
- *Events and sample space.* Some children only concentrate on a single event since their thinking is mainly deterministic (Langrall and Mooney 2005). It is then important that children understand the need to take into account all different possible outcomes in an experiment to compute its probability.
- *Combinatorial enumeration and counting.* Combinatorics is used in listing all the events in a sample space or in counting (without listing) all its elements. Although in the frequentist approach we do not need combinatorics to estimate the value of probability, combinatorial reasoning is nevertheless needed in other situations, for example, to understand how events in a compound experiment are formed or to understand how different samples of the same size can be selected from a population. Combinatorial reasoning is difficult; however, it is possible to use tools such as tree diagrams to help students reinforce this particular type of reasoning.
- *Independence and conditional probability.* The notion of independence is important to understand simulations and the empirical estimates of probability via frequency, since when repeating experiments we require independence of trials. Computing probabilities in compound experiments requires one to analyse whether the experiments are dependent or not. Finally, the idea of conditional probability is needed to understand many concepts in probability and statistics, such as confidence intervals or hypotheses tests.
- *Probability distribution and expectation.* Although there is abundant research related to distribution, most of this research concentrates on data distribution or in sampling distribution. Another type of distribution is linked to the random variable, a powerful idea in probability, as well as the associated idea of expectation. Some probability distribution models in wide use are the binomial, uniform, and normal distributions.
- *Convergence and laws of large numbers.* The progressive stabilization of the relative frequency of a given outcome in a large number of trials has been observed for centuries; Bernoulli proved the first version of the law of large numbers that justified the frequentist definition of probability. Today the frequentist approach, where probability is an estimate of the relative frequency of a result in a long series of trials, is promoted in teaching. It is important that students understand that each outcome is unpredictable and that regularity is only achieved in the long run. At the same time, older students should be able to discriminate between a frequency estimate (a value that varies) and probability (which is always a theoretical value) (Chaput et al. 2011).
- *Sampling and sampling distribution.* Given that we are rarely able to study complete populations, our knowledge of a population is based on samples. Students are required to understand the ideas of sample representativeness and sampling variability. The sampling distributions describe the variation of a summary

measure (e.g., sample means) along different samples from the same population. Instead of using the exact sampling distribution (e.g., a normal curve),
teaching currently favours the use of simulation or re-sampling to find an
empirical sampling distribution. This is a suitable teaching strategy, but teachers
should be conscious that, as any estimate, the empirical sampling distribution
only approximates the theoretical sampling distribution.
- *Modelling and simulation*. Today we witness increasing recommendations
  to approach the teaching of probability from the point of view of modelling
  (Chaput et al. 2011; Eichler and Vogel 2014; Prodromou 2014). Simulation
  allows the exploration of probability concepts and properties, and is used in
  informal approaches to inference. Simulation acts as an intermediary step
  between reality and the mathematical model. As a didactic tool, it can serve to
  improve students' probabilistic intuition, to acquire experience in the work of
  modelling, and to help students discriminate between model and reality.

2.4 Intuitions and Learning Difficulties

When teaching probability it is important to take into account the informal ideas
that children and adolescents assign to chance and probability before instruction.
These ideas are described in the breadth and depth of research investigating probabilistic intuitions, informal notions of probability, and resulting learning difficulties. Topics associated with teaching and learning probability, such as intuition,
informal notions of probability, cognition, misconceptions, heuristics, knowledge,
learning, reasoning, teaching, thinking, and understanding (among others), have
developed over the last 60 years of research investigating probabilistic thinking
(Chernoff and Sriraman 2014a).

We now revisit the essentials associated with probabilistic intuition and difficulties associated with learning probability. Framing our historical approach, we
adopt the periods of Jones and Thornton's (2005) "historical overview of research
on the learning and teaching of probability" (p. 66), which was recently extended
by Chernoff and Sriraman (2014b).

2.4.1 Piagetian and Post-piagetian Periods

Initial research in probability cognition was undertaken during the 1950s and 1960s
by Piaget and Inhelder and by psychologists with varying theoretical orientations
(Jones and Thornton 2005). Research during this period was dominated by the work
of Piaget and Inhelder,[4] which largely investigated "developmental growth and
structure of people's probabilistic thinking" (p. 65). Extensive investigations would

---

[4]To such an extent that Jones and Thornton denoted this period of research as "Piagetian Period."

reveal that children in particular stages were prone to subjective intuitions (Piaget and Inhelder 1951/1975). Alternatively stated, research investigating intuition and learning difficulties was central at the beginnings of research in probabilistic thinking and would continue on into the next (historical) phase of research.

The next phase of research, the "Post-piagetian Period," would be dominated, on the one hand, by work of Efraim Fischbein and, on the other hand, by Daniel Kahneman and Amos Tversky (Jones and Thornton 2005). The work of Fischbein would continue the work of Piaget and Inhelder (i.e., a continued focus on probabilistic intuitions), but introduce and explicate a distinction between what he denoted as primary and secondary intuitions. More specifically, Fischbein's (1975) notion of *primary intuitions* (not associated with formal instruction) continued the work of Piaget and Inhelder's research, but he differentiated his work during this period by investigating *secondary intuitions* (associated with formal instruction) and, further, how primary intuitions would not necessarily be overcome but rather replaced by secondary intuitions.

2.4.2 Heuristics and Biases Program

As mentioned, other investigations involving intuition were occurring in the field of psychology during this period, using different terminology.[5] Kahneman and Tversky's original *heuristics and biases program* (see, for example, Kahneman et al. 1982), which became widely known (e.g., Kahneman 2011) and was revisited 20 years later (Gilovich et al. 2002), was, in essence, an in-depth investigation into intuitions (and learning difficulties).

Kahneman and Tversky's research program investigated strategies that people used to calculate probabilities (*heuristics*) and the resultant systematic errors (*biases*). Their research revealed numerous heuristics (e.g., representativeness, availability and adjustment, and anchoring) and resulting biases. Years after generalising the notion of heuristics (Kahneman and Frederick 2002), Kahneman (2011) noted the essence of intuitive heuristics: "when faced with a difficult question, we often answer an easier one instead, usually without noticing the substitution" (p. 12). This research program played a key role in shaping many other fields of research (see, for example, behavioural economics).

In the field of mathematics education, the research of Shaughnessy (1977, 1981) brought forth not only the theoretical ideas of Tversky and Kahneman, but also, in essence, research on probabilistic intuitions and learning difficulties. Although not explicitly deemed as intuitions and difficulties, work in this general area of research was conducted by a number of different individuals.

In particular, Falk's (1981) research investigated difficulties associated with learning probability, and Konold (1989, 1991) conducted research looking into informal conceptions of probability (most notably *the outcome approach*). As the

---

[5]The theoretical constructs adopted by Daniel Kahneman and Amos Tversky differed from those of Fischbein and colleagues (e.g., Fischbein et al. 1970) and Piaget and Inhelder.

Post-piagetian Period came to a close, the field of mathematics education began to see an increasing volume of research on intuitions and learning difficulties (e.g., Green 1979, 1983, 1989; Fischbein and Gazit 1984), which is well summarised and synthesised in Shaughnessy's (1992) extensive chapter on research in probability and statistics education. Moving from one period to the next, research into probabilistic intuitions and learning difficulties would come into its own during what Jones (2005) called *Phase Three*: *Contemporary Research*.

2.4.3 Contemporary Research

During this new phase there was, arguably, a major shift towards investigating curriculum and instruction, and the leadership of investigating probabilistic intuitions and learning difficulties was carried on by a particular group of researchers. Worthy of note, mathematics education researchers in this phase, as the case with Konold (1989, 1991) and Falk (1981) in the previous phase, began to develop their own theories, frameworks, and models associated with responses to a variety of probabilistic tasks.

These theories, frameworks, and models were developed during research that investigated a variety of topics in probability, which included (difficulties associated with): randomness (e.g., Batanero et al. 2014; Batanero and Serrano 1999; Falk and Konold 1997; Pratt 2000), sample space (Jones et al. 1997, 1999; Nunes et al. 2014), and probabilistic reasoning (Fischbein et al. 1991; Fischbein and Schnarch 1997; Lecoutre 1992; Konold et al. 1993). Worthy of note, the term *misconception*, which acted as the de facto terminology for a number of years, has more recently evolved to *preconceptions* and other variants, which are perhaps better aligned with other theories in the field of mathematics education. In line with the above, research developing theories, models, and frameworks associated with intuition and learning difficulties continued into the next phase of research, which Chernoff and Sriraman (2014a) have (prematurely) called the Assimilation Period.

After more than 50 years of research investigating probabilistic intuitions and difficulties associated with learning probability, the field of mathematics education has, in essence, come into its own, which is evidenced by the type of research being conducted and the manner in which results are presented. Gone are the early days where researchers were attempting to replicate research found in different fields, such as psychology (e.g., Shaughnessy 1977, 1992). With that said, researchers are attempting to import theories, models, and frameworks from other fields; however, researchers in the field of mathematics education are forging their own interpretations of results stemming from the intuitive nature and difficulties associated with probability thinking and the teaching and learning of probability. Theories, models, and frameworks such as *inadvertent metonymy* (Abrahamson 2008, 2009), *sample space partitions* (Chernoff 2009, 2012a, b), and others demonstrate that research into intuitions and difficulties continues in the field of mathematics education.

This does not mean, however, that the field does not continue to look to other domains of research to help better inform mathematics education. For example, recent investigations (e.g., Chernoff 2012a, b) have gone back to their proverbial roots and integrated recent developments to the heuristics and biases program, which attempts to deal with the "arrested development of the representativeness heuristic in the field of mathematics education" (Chernoff 2012a, p. 951). Similar investigations embracing research from other fields have opened the door to alternative views of heuristics, intuitions, and learning difficulties, such as in the work by Gigerenzer and the Adaptive Behavior and Cognition (ABC) Group at the Max Planck Institute for Human Development in Berlin (e.g., Gigerenzer et al. 2011).

Based on these developments, the field of mathematics education is starting to also develop particular research which is building upon and questioning certain aspects of probabilistic intuitions and learning difficulties. For example, Chernoff, in a recent string of studies (e.g., Chernoff 2012a, b; Chernoff and Zazkis 2011), has begun to establish that perhaps normatively incorrect responses to a variety of probabilistic tasks are best accounted for not by heuristics or informal reasoning, but rather participants' use of logical fallacies.

In considering how students reason about probability, advances in technology and other educational resources have allowed for another important area of research, as described in the next section.

2.5 Technology and Educational Resources

Many educational resources have been used to support probability education. Some of the most common resources include physical devices such as dice, coins, spinners, marbles in a bag, and a Galton board that help create game-like scenarios that involve chance (Nilsson 2014). These devices are often used to support a classical approach to probability for computing the probability of an event occurring a priori by examining the object and making assumptions about symmetry that often lead to equiprobable outcomes for a single trial. When used together (e.g., two coins, a die, and a four-section spinner) these devices can be used to explore compound events and conditional probabilities (e.g., Martignon and Krauss 2009). Organizational tools such as two-by-two tables and tree diagrams are also used to assist in enumerating sample spaces (Nunes et al. 2014) and computing probabilities and can serve as important educational resources for students.

Since physical devices can also be acted upon, curriculum resources and teachers have increased the use of experiments with these devices to induce chance events (e.g., by rolling, spinning, choosing, or dropping a marble), often using relatively small sample sizes and recording frequencies of events. These frequencies and relative frequencies are used as an estimate of probability in the frequentist perspective, then often compared to the a priori computed probability based on examination of the object. Research by Nilsson (2007, 2009), for example, provides insights into students' thinking when engaged with experiments with such physical devices. Depending on the teachers' or the students' perspective, these experiments may favour one estimate of probability over another and issues related to sample size and the law of large numbers, or the difference between frequency and probability may or may not be discussed (Stohl 2005).

Many researchers (e.g., Chaput et al. 2011; Eicher and Vogel 2014; Lee and Lee 2011; Pratt 2011; Pratt and Ainley 2014; Prodromou 2014), and some curricula (e.g., CCSSI 2010; Raoult 2013), have recently emphasised that probability be taught as a way to model real-world phenomena rather than merely as an abstract measurement of something unseen about a real physical object (e.g., measure of likelihood of landing on heads when a coin is tossed). The sentiment is expressed by Pratt (2011):

> Of course, if the modelling meaning of probability was stressed in the curriculum, it is debatable whether there is much advantage in maintaining the current emphasis on coins, spinners, dice and balls drawn from a bag. Perhaps, in days gone by when children played board games, there was some natural relevance in such contexts but, now that games take place in real time on screens, probability has much more relevance as a tool for modelling computer-based action and for simulating real-world events and phenomena (p. 892).

One way to help students use probability to model real-world phenomena is to engage the necessity to make a model explicit when using technology. In Biehler's (1991) insightful comments on how technology should be used in probability education, he recommended several ways technology could advance learning and teaching probability. Sampling, storing, organising, and analysing data generated from a probabilistic model are facilitated tremendously by technology. These recommendations have been used by many researchers and have recently been made explicit for recommendations for teachers by Lee and Lee (2011) and for researchers by Pratt and Ainley (2014) and Pratt et al. (2011a, b).

Accordingly, we still need to make substantial progress to investigate how students' learning of probability models can be supported by the affordances of technology tools that will be used to frame the remaining discussion in the next sections.

2.5.1 Replicating Trials and Storing Data

One major contribution of technology to the study of probability is the ability to generate a large sample of data very quickly, store data in raw form as a sequence of outputs or organised table of frequencies, and collapse data into various aggregate representations. At times, the number of trials to perform is dictated by a teacher/researcher because of a pedagogical goal in mind; however, at other times, the number of trials is left open to be chosen by students.

Several researchers have discussed what students notice about sample size, particularly when they are able to examine its impact on variability in data distributions (e.g., Lee and Lee 2009; Lee et al. 2010; Pratt 2000). The ability of technology tools such as Probability Explorer or Fathom to store long lists of data sequences can also afford opportunities for students to examine a history of outcomes in the order in which they occurred, as well as conveniently collapsed in frequency and relative frequency tables and graphs.

2.5.2 Representing and Analysing Data

Technology tools bring real power to classrooms by allowing students to rapidly generate a large amount of data, quickly create tabular and graphical representations, and perform computations on data with ease. Students can then spend more time focused on making sense of data in various representations. Different representations of data in aggregate form can afford different perspectives on a problem. In addition, technology facilitates quickly generating, storing, and comparing multiple samples, each consisting of as many trials as desired. Instead of having each student in a class collect a small amount of data and then pool the data to form a class aggregate, students (or small groups) can generate individual samples of data, both small and large, and engage in reasoning about their own data as well as contribute to class discussions that examine results across samples (Stohl and Tarr 2002).

Technology can also provide opportunities for students to make sense of how an empirical distribution changes *as data is being collected*. This dynamic view of a distribution can assist students in exploring the relationship between a probability model and a resulting empirical distribution and how sample size impacts the variability observed (Drier 2000; Pratt et al. 2011a, b). This relies upon intuitions about laws of large numbers; such intuitions may be strengthened by observing the settling down of relative frequency distributions as trials are simulated and displayed in real time. This type of work provides promising tasks and technology tools as well as future directions for how we can build on this work to better understand students' reasoning with such dynamic representations.

2.5.3 Probability Technological Models

It is important to investigate the opportunities that technology affords for teachers and students to discuss explicitly the assumptions needed to build models to describe real-world scenarios through simulation. The model-building process should include discussing the pertinent characteristics of a situation being modelled, while in the same way simplifying reality (Chaput et al. 2011; Eichler and Vogel 2014). In the next steps, creating a mathematical model and working with it, the student should find ways to partition the events in which the probabilities are easily identifiable, using physical "chance makers" to model the random processes if possible and building and working with a simulated model with technological tools. Such steps are opportunities for students to grow in their understandings of a situation, the model, and many probability ideas. Using various technology tools (e.g., a graphing calculator, spreadsheet, or software such as Fathom or TinkerPlots) to create simulations for the same situation can force students to consider carefully how different commands in a tool can be used to create a model for a situation. Further, this can afford opportunities for discussing why two different ways of modelling a situation may be similar or different and differentiating between model and reality.

The modelling process may be as "simple" as having students design a discrete probability distribution that models a real spinner with three unequal sectors (e.g., Stohl and Tarr 2002) or designing a working model to consider whether larger families tend to either have more girls or more boys rather than the same number of boys and girls (Lee and Lee 2011). Modelling may also be as complex as interpreting medical information using input probabilities from real-life scenarios such as success or complications from back surgery to help a patient make an important life decision (e.g., Pratt et al. 2011a, b).

Several tasks and tools discussed by Prodromou (2014) and Pratt and Ainley (2014) illustrate the importance of the ability to adjust parameters in a model-building and model-fitting process. The ability to easily adjust parameters in a model can afford many opportunities for students to explore "what if" scenarios. It allows for a process of model development, data generation and analysis, and model adjustment. Konold and Kazak (2008), for example, described how this process helped students in creating and modifying models based on how well a particular model behaved when a simulation was run.

2.5.4 Hiding a Model

Engaging with probability models and the data generated from such models can provide very important foundations for how probability is used in statistics, particularly in making inferences about populations and testing hypotheses. In the above cases, the models used in a simulation are typically created by students or created by a teacher but open for inspection by students. However, technology tools afford the ability to hide a model from the user such that underlying probability distributions that control a simulation are unknowable. These "black-box" types of simulations may assist students in thinking about probability from a subjective or frequentist perspective where they can only use data generated from a simulation to make estimates of probabilities that they can use in inference or decision-making situations. One example of such inference and decision-making situations can be found in the work of Lee et al. (2010) where 11- and 12-year-olds investigate whether a die company produces fair dice to make a decision about whether to buy dice from the company for a board game.

In summary, technology provides a big opportunity for probability education but also sets some challenges. One of them is the education of teachers to teach probability in general and to use technology in their teaching of probability in particular. We deal with this specific issue in the last section of our survey.

2.6 Education of Teachers

A consequence of the philosophical debates around the meaning of probability, the particular features of probabilistic reasoning, the students' misconceptions and difficulties, and the increasing variety of technological resources is that teachers need specific preparation to teach probability. Although school textbooks provide examples and teaching resources, some texts present too narrow a view of probabilistic concepts or only one approach to probability. The applications of probability in textbooks may be restricted to games of chance and/or the definitions of concepts may be incorrect or incomplete (Cañizares et al. 2002).

Research in this area is scarce and is mostly focussed on the evaluation of prospective teachers' knowledge of probability (e.g., Batanero et al. 2014; Chernoff and Russel 2012). While there is still need for further research into prospective and practising teacher probability knowledge, two important missing areas of research in probability education are the analysis of the components of teachers' knowledge and the design of adequate materials and effective activities for educating teachers.

2.6.1 Components in Teacher Knowledge

Existent models of the knowledge needed by teachers in mathematics education, such as mathematical knowledge for teaching (MKT; Ball et al. 2008), suggest that teachers need different types of content and pedagogical knowledge. However, as stated by Godino et al. (2011) and Groth and Bergner (2013), for the particular case of statistical knowledge for teaching (SKT) it is important to recognise that teachers need content-specific knowledge to guide instruction. This means that any discussion of probability knowledge for teaching (PKT) should be supported in the specific features of probability.

First, teachers need adequate probabilistic knowledge. However, even if prospective teachers have a degree in mathematics, they have usually only studied theoretical probability and lack experience in designing investigations or simulations to work with students (Kvatinsky and Even 2002; Stohl 2005). The education of primary school teachers is even more challenging, because few of them have had suitable training in either theoretical or applied probability (Franklin and Mewborn 2006). Moreover, recent research suggests that many prospective teachers share with their students common biases in probabilistic reasoning (e.g., Batanero et al. 2014; Prodromou 2012)

A second component is the pedagogical knowledge needed to teach probability, where general principles valid for other areas of mathematics are not always appropriate (Batanero et al. 2004). For example, in arithmetic or geometry, elementary operations can be reversed, and this reversibility can be represented by concrete materials, which serve to organise experiences where children progressively abstract the structure behind the concrete situation. The lack of reversibility in random experiences makes it more difficult for children to grasp the essential features of randomness, which may explain why they do not always develop correct probabilistic intuitions without a specific instruction.

In addition to the above, probability is difficult to teach because the teacher should not only present different probabilistic concepts and their applications but be aware of the different meanings of probability and philosophical controversies around them (Batanero et al. 2004). Finally, teachers should be acquainted with research results that describe children's reasoning and beliefs in uncertain situations and with didactic materials that can help their students develop correct intuitions in this field.

The current use of technology warrants special considerations in the education of teachers that should be analysed. Lee and Hollebrands (2011) introduced a framework to describe what they call technological pedagogical statistical knowledge (TPSK) with examples of components in this knowledge. The evaluation and development of components in this framework for the specific case of probability is a promising research area.

2.6.2 Effective Ways to Train Teachers

Another line of research is designing and evaluating suitable and effective tasks that help in increasing the probabilistic and didactic knowledge of teachers. Some researchers describe different experiences directed towards achieving this goal.

Teachers should engage with and analyse probability simulations and investigations. Simulations and experiments are recommended when working with students. To be able to use investigations in their own classrooms, teachers need competencies with this approach to teaching. When the time available for educating teachers is scarce, one possibility is to give teachers first a project or investigation to work with and, when finished, carry out a didactical analysis of the project. This type of analysis can help to simultaneously increase the teachers' mathematical and pedagogical knowledge (Batanero et al. 2004).

Teachers should engage with case discussions. Groth and Xu (2011) used case discussion among a group of teachers as a valuable strategy to educate teachers. The authors indicated that in teaching stochastics teachers navigate between two layers of uncertainty. On the one hand, uncertainty is part of stochastic knowledge; on the other hand, in any classroom uncertainty appears as a result of the dynamic interactions amongst teacher, students, and the topic being taught. Discussions among the teachers may help them to increase their knowledge since experiences with general pedagogy, mathematical content, and content-specific pedagogy can be offered and debated.

Teachers also need experience planning and analysing a lesson. When teachers plan and then analyse a lesson devised to teach some content they develop their probabilistic and professional knowledge (Chick and Pierce 2008). Teachers need to understand the probability they teach to their students. One strategy is to have teachers play the role of a learner and afterwards analyse what they learnt. If they have the chance to go through a lesson as a learner and at the same time look at it from the point a view of a teacher, they may understand better how the lesson will unfold later in the classroom.

Teachers should have extensive experience working with technology. We can also capitalise on technology as a tool-builder for teachers gaining a conceptual understanding of probabilistic ideas. Lee and Hollebrands (2011) describe the way technology can function both as amplifiers and reorganisers of teachers' knowledge. They also discuss how technology can provide teachers with first-hand experience about how these tools can be useful in improving their stochastic thinking and knowledge. Other examples describe experiences and courses specifically directed to train teachers to teach probability or suggestions of how this training should be (e.g., Batanero et al. 2005a, b; Dugdale 2001; Kvatinsky and Even 2002; Stohl 2005).

Research and development in teacher education related to probability education is still scarce and needs to be fostered.

3 Summary and Looking Ahead

In the previous sections we analysed the multifaceted nature of probability, the probabilistic contents in curricula, research dealing with intuitions and misconceptions, the role of technology, and the education of teachers. To finish this survey we suggest some points where new research is needed.

Different views on probability: As discussed in Sect. 2.1, the different views of probability are linked to philosophical debates, and the school curricula at times has given too much emphasis to only one probability view. Since different views of probability are complementary (Henry 2010), reducing teaching to just one approach may explain some learning difficulties, as students may consider or apply only one interpretation in situations where it is inappropriate. Some research questions on this topic include: (a) an analysis of the particular views of probability implicit in curricular documents and textbooks for different school levels, (b) an exploration and consequent recommendations about the best age at which a particular view of probability should be introduced to students and about the best sequence to introduce different approaches to probability, (c) the design and evaluation of curricular guidelines for different ages that take into account each particular view of probability, and (d) the analyses of teachers' educational needs in relation to each particular view of probability.

Probabilistic thinking and reasoning: Our analysis in Sect. 2.2 suggests that probabilistic reasoning complements logical, causal, and statistical reasoning. Consequently, the teaching of probabilistic thinking is important and justified in its own right and not simply as a tool to pave the way to inferential methods of statistics. Important research problems in this regard are: (a) clarifying the way in which probabilistic thinking could contribute to improving mathematical competencies of students, (b) analysing how different probability models and their applications can be presented to the students, (c) finding ways in which it is possible to engage students in questions related to how to obtain knowledge from data and why a probability model is suitable, and (d) how to help students develop valid intuitions in this field.

Probability in school curricula: Another important area of research is to enquire about how the fundamental ideas of probability have been reflected in school curricula at different levels and in different countries. The presentation of these ideas in textbooks for different curricular levels should also been taken into account (following previous research, e.g., Azcárate et al. 2006; Jones and Tarr 2007). We also need to find different levels of formalisation to teach each of these ideas depending on age and previous knowledge of students. Consequently, it is important to reflect on the main ideas that students should acquire at different ages, appropriate teaching methods, and suitable teaching situations.

Students' intuitions and learning difficulties: As suggested in previous research, probabilistic intuitions and difficulties in learning probability, albeit described in a variety of different forms, are a mainstay of research in psychology and mathematics education. Further, as the field grows and diversifies, there is reason to expect that this particular thread of research will not only continue but also grow. Important research questions in this area are: (a) describing primary school children's preconceptions and intuitions in relation to new probability content in school curricula, (b) analysing changes in students' intuitions about different probability concepts after specific teaching experiments, and (c) how "best" to account for preconceptions and difficulties.

Technology and education of teachers: The quick changes in technological development suggest a need for new research on students' use of technology to solve real problem scenarios that use probability models. We need to know more about how students construct models and how they reason with data generated from such models. It is also important to evaluate the impact of technology on recent curricula and on the education of teachers. There is also a need for more systematic research about how teachers and students use technology in classrooms and how large-scale assessment should respond to capture new meanings for probability that may emerge from students working with probability using technology tools.

In conclusion, in this brief survey we have tried to summarise the extensive research in probability education. At the same time we intended to convince our readers of the need for new research and the many different ideas that still need to be investigated. We hope to have achieved these goals and look forward to new research in probability education.

Reference

Abrahamson, D. (2008). Bridging theory: Activities designed to support the grounding of outcome-based combinatorial analysis in event-based intuitive judgment-A case study. In M. Borovcnik & D. Pratt (Eds.), *Proceedings of Topic Study Group 13 at the 11th International Conference on Mathematics Education (ICME)*. Monterrey, Mexico. http://edrl.berkeley.edu/pubs/Abrahamson-ICME11-TSG13_BridgingTheory.pdf.

Abrahamson, D. (2009). Orchestrating semiotic leaps from tacit to cultural quantitative reasoning: the case of anticipating experimental outcomes of a quasi-binomial random generator. *Cognition and Instruction, 27*(3), 175–224.

Australian Curriculum, Assessment and Reporting Authority (ACARA) (2010). *The Australian curriculum: Mathematics*. Sidney, NSW: Author. http://www.australiancurriculum.edu.au/mathematics/curriculum/f-10?layout=1.

Azcárate, P., Cardeñoso, J. M., & Serradó, S. (2006). Randomness in textbooks: the influence of deterministic thinking. In M. Bosch (Ed.), *Proceedings of the Fourth Conference of the European Society for Research in Mathematics Education*. Sant Feliu de Guixols, Spain: ERME. http://fractus.uson.mx/Papers/CERME4/Papers%20definitius/5/SerradAzcarCarde.pdf.

Ball, D. L., Thames, M. H., & Phelps, G. (2008). Content knowledge for teaching. *Journal of Teacher Education, 59*(5), 389–407.

Batanero, C. (2013). Teaching and learning probability. In S. Lerman (Ed.), *Encyclopedia of mathematics education* (pp. 491–496). Heidelberg: Springer.

Batanero, C. (2015). Understanding randomness: Challenges for research and teaching. Plenary lecture. *Ninth European Conference of Mathematics Education*. Prague, Czech Republic.

Batanero, C., Arteaga, P., Serrano, L., & Ruiz, B. (2014). Prospective primary school teachers' perception of randomness. In E. Chernoff & B. Sriraman (Eds.), *Probabilistic thinking: Presenting plural perspectives* (pp. 345–366). New York: Springer.

Batanero, C., Biehler, R., Maxara, C., Engel, J., & Vogel, M. (2005a). Using simulation to bridge teachers' content and pedagogical knowledge in probability. *Paper presented at the fifteenth ICMI Study Conference: The professional education and development of teachers of mathematics*. Aguas de Lindoia, Brazil: International Commission for Mathematical Instruction.

Batanero, C., & Díaz, C. (2007). Meaning and understanding of mathematics. The case of probability. In J. P Van Bendegen & K. François (Eds), *Philosophical dimensions in mathematics education* (pp. 107–127). New York: Springer.

Batanero, C., Godino, J. D., & Roa, R. (2004). Training teachers to teach probability. *Journal of Statistics Education, 12*. http://www.amstat.org/publications/jse/v12n1/batanero.html.

Batanero, C., Henry, M., & Parzysz, B. (2005b). The nature of chance and probability. In G. A. Jones (Ed.), *Exploring probability in school: challenges for teaching and learning* (pp. 15–37). New York: Springer.

Batanero, C., & Serrano, L. (1999). The meaning of randomness for secondary school students. *Journal for Research in Mathematics Education, 30*(5), 558–567.

Bennett, D. J. (1999). *Randomness*. Cambridge, MA: Harvard University Press.

Ben-Zvi, D., & Garfield, J. B. (Eds.). (2004). *The challenge of developing statistical literacy, reasoning and thinking*. Dordrecht, The Netherlands: Kluwer.

Bernoulli, J. (1987). *Ars conjectandi*, Rouen: IREM. (Original work published in 1713).

Biehler, R. (1991). Computers in probability education. In R. Kapadia & M. Borovcnik (Eds.), *Chance encounters: Probability in education* (pp. 169–211). Dordrecht, The Netherlands: Kluwer.

Borovcnik, M. (2005). Probabilistic and statistical thinking, In M. Bosch (Ed.), *Proceedings of the Fourth Conference on European Research in Mathematics Education*. Sant Feliu de Guissols, Spain: ERME. http://fractus.uson.mx/Papers/CERME4/Papers%20definitius/5/Borovcnik.pdf.

Borovcnik, M. (2011). Strengthening the role of probability within statistics curricula. In C. Batanero, G. Burrill, & C. Reading (Eds.) (2011). *Teaching Statistics in School Mathematics- Challenges for Teaching and Teacher Education. A Joint ICMI/IASE Study* (pp. 71–83). New York: Springer.

Borovcnik, M. (2012). Multiple perspectives on the concept of conditional probability. *Avances de Investigación en Educación Matemática, 2,* 5–27. http://www.aiem.es/index.php/aiem/article/view/32.

Borovcnik, M., Bentz, H. J., & Kapadia, R. (1991). Empirical research in understanding probability. In R. Kapadia & M. Borovcnik (Eds.), *Chance encounters: Probability in education* (pp. 73–105). Dordrecht, The Netherlands: Kluwer.

Borovcnik, M., & Kapadia, R. (2014a). A historical and philosophical perspective on probability. In E. J Chernoff & B. Sriraman, (Eds.), *Probabilistic thinking: presenting plural* perspectives (pp. 7–34). New York: Springer.

Borovcnik, M., & Kapadia, R. (2014b). From puzzles and paradoxes to concepts in probability. In E. J. Chernoff & B. Sriraman (Eds.), *Probabilistic thinking: presenting plural perspectives* (pp. 35–73). New York: Springer.

Cañizares, M. J., Ortiz, J. J., Batanero, C., & Serrano, L. (2002). Probabilistic language in Spanish textbooks. In B. Phillips (Ed.), *ICOTS-6 papers for school teachers* (pp. 207–211). Cape Town: International Association for Statistical Education.

Cardano, G. (1961). *The book on games of chances.* New York: Holt, Rinehart & Winston (Original work published in 1663).

Carnap, R. (1950). *Logical foundations of probability.* Chicago: University of Chicago Press.

Chaput, B., Girard, J. C., & Henry, M. (2011). Frequentist approach: Modelling and simulation in statistics and probability teaching. In C. Batanero, G. Burrill, & C. Reading (Eds.), *Teaching Statistics in school mathematics-challenges for teaching and teacher education* (pp. 85–95). New York: Springer.

Chernoff, E. J. (2009). Sample space partitions: An investigative lens. *Journal of Mathematical Behavior, 28*(1), 19–29.

Chernoff, E. J. (2012a). Logically fallacious relative likelihood comparisons: the fallacy of composition. *Experiments in Education, 40*(4), 77–84.

Chernoff, E. J. (2012b). Recognizing revisitation of the representativeness heuristic: an analysis of answer key attributes. *ZDM - The International Journal on Mathematics Education, 44*(7), 941–952.

Chernoff, E. J., & Russell, G. L. (2012). The fallacy of composition: Prospective mathematics teachers' use of logical fallacies. *Canadian Journal of Science, Mathematics and Technology Education, 12*(3), 259–271.

Chernoff, E. J. & Sriraman, B. (2014a). Introduction. In E. J. Chernoff & B. Sriraman (Eds.), *Probabilistic thinking: presenting plural perspectives* (pp. xv–xviii). New York: Springer.

Chernoff, E. J. & Sriraman, B. (2014b). Commentary on probabilistic thinking: presenting plural perspectives. In E. J. Chernoff & B. Sriraman (Eds.), *Probabilistic thinking: presenting plural perspectives* (pp. 721–728). New York: Springer.

Chernoff, E. J., & Zazkis, R. (2011). From personal to conventional probabilities: from sample set to sample space. *Educational Studies in Mathematics, 77*(1), 15–33.

Chick, H. L., & Pierce, R. U. (2008). Teaching statistics at the primary school level: beliefs, affordances, and pedagogical content knowledge. In C. Batanero, G. Burrill, C. Reading, & A. Rossman (Eds.), *Proceedings of the ICMI study 18 and IASE round table conference.* International Commission on Mathematics Instruction and International Association for Statistical Education: Monterrey, Mexico.

Chung, K. L. (2001). *A course in probability theory.* London: Academic Press.

Common Core State Standards Initiative (CCSSI). (2010). *Common Core State Standards for Mathematics.* Washington, DC: National Governors Association for Best Practices and the Council of Chief State School Officers. http://www.corestandards.org/Math/.

David, F. N. (1962). *Games, gods and gambling.* London: Griffin.

de Finetti, B. (1933). Sul concetto di probabilità [On the concept of probability]. *Rivista Italiana di Statistica, Economia e Finanza, 5,* 723–747.

de Moivre, A. (1967). *The doctrine of chances.* New York: Chelsea (Original work published in 1718).

Drier, H. S. (2000). Children's meaning-making activity with dynamic multiple representations in a probability microworld. In M. Fernandez (Ed.), *Proceedings of the twenty-second annual meeting of the North American Chapter of the International Group for the Psychology of Mathematics Education* (Vol. 2, pp. 691–696). Tucson, AZ: North American Chapter of the International Group for the Psychology of Mathematics Education.

Dugdale, S. (2001). Pre-service teachers' use of computer simulation to explore probability. *Computers in the Schools, 17*(1–2), 173–182.

Eichler, A., & Vogel, M. (2014). Three approaches for modelling situations with randomness. In E. J. Chernoff & B. Sriraman (Eds.), *Probabilistic thinking: Presenting plural perspectives* (pp. 75–99). New York: Springer.

Engel, J., & Sedlmeier, P. (2005). On middle-school students' comprehension of randomness and chance variability in data-. *Zentralblatt Didaktik der Mathematik, 37*(3), 168–177.

Engel, J., Sedlmeier, P., & Worn, C. (2008). Modelling scatterplot data and the signal-noise metaphor: Towards statistical literacy for pre-service teachers. In C. Batanero, G. Burrill, C. Reading, & A. Rossman (Eds.), *Proceedings of the ICMI study 18 and IASE round table conference*. International Commission on Mathematics Instruction and International Association for Statistical Education: Monterrey, Mexico.

Falk, R. (1981). The perception of randomness. *Proceedings of the fifth conference of the International Group for the Psychology of Mathematics Education* (pp. 222–229). Grenoble, France: University of Grenoble.

Falk, R., & Konold, C. (1992). The psychology of learning probability. In F. S. Gordon & S. P. Gordon (Eds.), *Statistics for the twenty-first century* (pp. 151–164). Washington: Mathematical Association of America.

Falk, R., & Konold, C. (1997). Making sense of randomness: Implicit encoding as a basis for judgement. *Psychological Review, 104*(2), 310–318.

Fine, T. L. (1971). *Theories of probability. An examination of foundations*. London: Academic Press.

Fischbein, E. (1975). *The intuitive source of probability thinking in children*. Dordrecht, The Netherlands: Reidel.

Fischbein, E., & Gazit, A. (1984). Does the teaching of probability improve probabilistic intuitions? *Educational Studies in Mathematics, 15*, 1–24.

Fischbein, E., Nello, M. S., & Marino, M. S. (1991). Factors affecting probabilistic judgments in children and adolescents. *Educational Studies in Mathematics, 22*, 523–549.

Fischbein, E., Pampu, I., & Minzat, I. (1970). Comparison of ratios and the chance concept in children. *Child Development, 41*, 377–389.

Fischbein, E., & Schnarch, D. (1997). The evolution with age of probabilistic, intuitively based misconceptions. *Journal for Research in Mathematics Education, 28*, 96–105.

Franklin, C., & Mewborn, D. (2006). The statistical education of PreK-12 teachers: A shared responsibility. In G. Burrill (Ed.), *NCTM 2006 Yearbook: Thinking and reasoning with data and chance* (pp. 335–344). Reston, VA: National Council of Teachers of Mathematics.

Franklin, C., Kader, G., Mewborn, D., Moreno, J., Peck, R., Perry, M., et al. (2007). *Guidelines for assessment and instruction in statistics education (GAISE) report: A Pre-K-12 curriculum framework*. Alexandria, VA: American Statistical Association. http://www.amstat.org/Education/gaise/.

Gal, I. (2005). Towards "probability literacy" for all citizens: Building blocks and instructional dilemmas. In G. A. Jones (Ed.), *Exploring probability in school. Challenges for teaching and learning* (pp. 39–63). Dordrecht, The Netherlands: Kluwer.

Gigerenzer, G., Hertwig, R., & Pachur, T. (2011). *Heuristics: The Foundations of Adaptive Behavior*. Oxford, MA: University Press.

Gillies, D. (2000). Varieties of propensities. *British Journal of Philosophy of Science, 51*, 807–835.

Gilovich, T., Griffin, D., & Kahneman, D. (2002). *Heuristics and biases: The psychology of intuitive judgment*. New York: Cambridge University Press.

Godino, J. D., Ortiz, J. J., Roa, R., & Wilhelmi, M. R. (2011). Models for statistical pedagogical knowledge. In C. Batanero, G. Burrill, & C. Reading (Eds.), *Teaching statistics in school mathematics-challenges for teaching and teacher education* (pp. 271–282). New York: Springer.

Green, D. (1979). The chance and probability concepts project. *Teaching Statistics, 1*(3), 66–71.

Green, D. (1983). School pupils' probability concepts. *Teaching Statistics, 5*(2), 34–42.

Green, D. R. (1989). Schools students' understanding of randomness. In R. Morris (Ed.), *Studies in mathematics education: The teaching of statistics* (Vol. 7, pp. 27–39). Paris: UNESCO.

Groth, R. E., & Bergner, J. S. (2013). Mapping the structure of knowledge for teaching nominal categorical data analysis. *Educational Studies in Mathematics, 83*, 247–265.

Groth, R. E., & Xu, S. (2011). Preparing teachers through case analyses. In C. Batanero, G. Burrill, & C. Reading (Eds.), *Teaching statistics in school mathematics-challenges for teaching and teacher education* (pp. 371–382). New York: Springer.

Hacking, I. (1975). *The emergence of probability*. Cambridge, MA: Cambridge University Press.

Heitele, D. (1975). An epistemological view on fundamental stochastic ideas. *Educational Studies in Mathematics, 6*, 187–205.

Henry, M. (2010). Evolution de l'enseignement secondaire français en statistique et probabilities [Evolution of French secondary teaching in statistics and probability]. *Statistique et Enseignement, 1*(1), 35–45.

Jones, J. L., & Tarr, J. E. (2007). An examination of the levels of cognitive demand required by probability tasks in middle grade mathematics textbooks. *Statistics Education Research Journal, 6*(2), 4–27.

Jones, G. A., Langrall, C. W., Thornton, C. A., & Mogill, A. T. (1997). A framework for assessing and nurturing young children's thinking in instruction. *Educational Studies in Mathematics, 32*, 101–125.

Jones, G. A., Langrall, C. W., Thornton, C. A., & Mogill, A. T. (1999). Students' probabilistic thinking in instruction. *Journal for Research in Mathematics Education, 30*, 487–519.

Kahneman, D. (2011). *Thinking fast and slow*. New York: MacMillan.

Kahneman, D., Slovic, P., & Tversky, A. (1982). *Judgment under uncertainty: Heuristics and biases*. New York: Cambridge University Press.

Kahneman, D., & Frederick, S. (2002). Representativeness revisited: attribute substitution in intuitive judgement. In T. Gilovich, D. Griffin, & D. Kahneman (Eds.), *Heuristics and biases: The psychology of intuitive judgment* (pp. 49–81). New York: Cambridge University Press.

Keynes, J. M. (1921). *A treatise on probability*. New York: MacMillan.

Kolmogorov, A. (1950). *Foundations of probability's calculation*. New York: Chelsea Publishing Company (Original work, published in 1933).

Konold, C. (1989). Informal conceptions of probability. *Cognition and Instruction, 6*, 59–98.

Konold, C. (1991). Understanding students' beliefs about probability. In E. Von Glasersfeld (Ed.), *Radical constructivism in mathematics education* (pp. 139–156). Dordrecht, The Netherlands: Kluwer.

Konold, C., & Kazak, S. (2008). Reconnecting data and chance. *Technology Innovations in Statistics Education, 2*(1). https://escholarship.org/uc/item/38p7c94v.

Konold, C., & Pollatsek, A. (2002). Data analysis as the search for signals in noisy processes. *Journal for Research in Mathematics Education, 33*(4), 259–289.

Konold, C., Pollatsek, A., Well, A., Lohmeier, J., & Lipson, A. (1993). Inconsistencies in students' reasoning about probability. *Journal for Research in Mathematics Education, 24*(5), 392–414.

Kultusministerkonferenz (KMK). (2004). *Bildungsstandards im Fach Mathematik für den mittleren Schulabschluss* [Educational standards in mathematics for middle school]. Berlin: Author.

Kultusministerkonferenz (KMK) (2012). *Bildungsstandards im Fach Mathematik für die Allgemeine Hochschulreife* [Educational standards in mathematics for the general higher education]. Berlin: Author.

Kvatinsky, T., & Even, R. (2002). Framework for teacher knowledge and understanding of probability. In B. Phillips (Ed.), *Proceedings of the sixth international conference on teaching statistics*. International Statistical Institute: Voorburg, The Netherlands.

Laplace, P. S. (1986). *Essai philosophique sur les probabilités* [Philosophical essay on Probabilities]. Paris: Christian Bourgois (Original work, published in 1814).

Langrall, C. W., & Mooney, E. S. (2005). Characteristics of elementary school students' probabilistic reasoning. In G. Jones (Ed.), *Exploring probability in school* (pp. 95–119). New York: Springer.

Lee, H. S., Angotti, R. L., & Tarr, J. E. (2010). Making comparisons between observed data and expected outcomes: Students' informal hypothesis testing with probability simulation tools. *Statistics Education Research Journal, 9*(1), 68–96.

Lee, H. S., & Hollebrands, K. F. (2011). Characterising and developing teachers' knowledge for teaching statistics with technology. In C. Batanero, G. Burrill, & C. Reading (Eds.), *Teaching statistics in school mathematics-challenges for teaching and teacher education* (pp. 359–369). Netherlands: Springer.

Lee, H. S., & Lee, J. T. (2009). Reasoning about probabilistic phenomena: Lessons learned and applied in software design. *Technology Innovations in Statistics Education 3*(2). https://escholarship.org/uc/item/1b54h9s9.

Lee, H. S., & Lee, J. T. (2011). Simulations as a path for making sense of probability. In K. Hollebrands & T. Dick (Eds.), *Focus in high school mathematics on reasoning and sense making with technology* (pp. 69–88). Reston, VA: National Council of Teachers of Mathematics.

Lecoutre, M. (1992). Cognitive models and problem spaces in "purely random" situations. *Educational Studies in Mathematics, 23*, 557–568.

Martignon, L. (2014). Fostering children's probabilistic reasoning and first elements of risk evaluation In E. J. Chernoff, B. & Sriraman (Eds.), *Probabilistic thinking, presenting plural perspectives* (pp. 149–160). Dordrecht: The Netherlands: Springer.

Martignon, L., & Krauss, S. (2009). Hands on activities with fourth-graders: a tool box of heuristics for decision making and reckoning with risk. *International Electronic Journal for Mathematics Education, 4*, 117–148.

Mellor, D. H. (1971). *The matter of chance*. Cambridge: Cambridge University Press.

Ministerio de Educación y Ciencia, MEC. (2006). *Real Decreto 1513/2006, de 7 de diciembre, por el que se establecen las enseñanzas mínimas de la Educación Primaria* [Royal Decree establishing the minimum content for Primary Education] Madrid: Author.

Ministerio de Educación y Ciencia, MEC. (2007a). *Real Decreto 1631/2006, de 29 de diciembre, por el que se establecen las enseñanzas mínimas correspondientes a la Educación Secundaria Obligatoria* [Royal Decree establishing the minimum content for Compulsory Secondary Education]. Madrid: Author.

Ministerio de Educación y Ciencia, MEC. (2007b). *Real Decreto 1467/2007, de 2 de noviembre, por el que se establece la estructura del bachillerato y se fijan sus enseñanzas mínimas* [Royal Decree establishing the structure and minimum content for High School]. Madrid: Author.

Ministerio de Educación, Cultura y Deporte, MECD. (2014). *Real Decreto 126/2014, de 28 de febrero, por el que se establece el currículo básico de la Educación Primaria* [Royal Decree establishing the minimum content for Primary Education]. Madrid: Author.

Ministerio de Educación, Cultura y Deporte, MECD. (2015). *Real Decreto 1105/2014, de 26 de diciembre, por el que se establece el currículo básico de la Educación Secundaria Obligatoria y del Bachillerato* [Royal Decree establishing the minimum content for Compulsory Secondary Education and High School]. Madrid: Author.

Ministerio de Educación Pública, MEP. (2012). *Programas de Estudio de Matemáticas* [Study programs for mathematics]. San José: Costa Rica: Author.

Ministry of Education, ME. (2007). *The New Zealand curriculum*. Wellington, New Zealand: Learning Media.

Moore, D. S. (2010). *The basic practice of statistics*. New York: Freeman (5th edition).

National Council of Teachers of Mathematics, NCTM. (2000). *Principles and standards for school mathematics*. Reston, VA: Author.

Nilsson, P. (2007). Different ways in which students handle chance encounters in the explorative setting of a dice game. *Educational Studies in Mathematics, 66*, 293–315.

Nilsson, P. (2009). Conceptual variation and coordination in probability reasoning. *The Journal of Mathematical Behavior, 28*(4), 247–261.

Nilsson, P. (2014). Experimentation in probability teaching and learning. In E. Chernoff & B. Sriraman (Eds.), *Probabilistic thinking. Presenting multiple perspectives* (pp. 509–532). New York: Springer.

Nunes, T., Bryant, P., Evans, D., Gottardis, L., & Terlektsi, M. E. (2014). The cognitive demands of understanding the sample space. *ZDM - The International Journal on Mathematics Education, 46*(3), 437–448.

Pange, J., & Talbot, M. (2003). Literature survey and children's perception on risk. *ZDM - The International Journal on Mathematics Education, 35*(4), 182–186.

Pascal, B. (1963). Correspondance avec Fermat [Correspondence with Fermat]. In B. Pascal, *Oeuvres Complètes* (pp. 43–49). París: Seuil (Original letter written in 1654).

Peirce, C. S. (1932). Notes on the doctrine of chances. In C. S. Peirce, *Collected papers* (Vol. 2, pp. 404–414). Havard University Press (Original work published in 1910).

Piaget, J., & Inhelder, B. (1975). *The origin of the idea of chance in children*. New York: Norton. (Original work published in 1951).

Popper, K. R. (1959). The propensity interpretation of probability. *British Journal of the Philosophy of Science, 10*, 25–42.

Pratt, D. (2000). Making sense of the total of two dice. *Journal for Research in mathematics Education, 31*(5), 602–625.

Pratt, D. (2011). Re-connecting probability and reasoning about data in secondary school teaching. *Paper presented at 58th ISI World Statistics Congress*, Dublin, Ireland. http://2011.isiproceedings.org/papers/450478.pdf.

Pratt, D., & Ainley, J. (2014). Chance re-encounters: Computers in probability education revisited. In T. Wassong (Ed.), *Mit Werkzeugen mathematik und stochastik lernen–using tools for learning mathematics and statistics* (pp. 165–177). Wiesbaden: Springer Fachmedien.

Pratt, D., Ainley, J., Kent, P., Levinson, R., Yogui, C., & Kapadia, R. (2011a). Role of context in risk-based reasoning. *Mathematical Thinking and Learning, 13* (4), 322–345.

Pratt, D., Davies, N., & Connor, D. (2011b). The role of technology in teaching and learning statistics. In C. Batanero, G. Burrill, & C. Reading (Eds.). *Teaching statistics in school mathematics-challenges for teaching and teacher education* (pp. 97–107). New York: Springer.

Prodromou, T. (2012). Connecting experimental probability and theoretical probability. *ZDM - The International Journal on Mathematics Education, 44*(7), 855–868.

Prodromou, T. (2014). Developing a modelling approach to probability Using computer-based simulations. In E. Chernoff & B. Sriraman (Eds.), *Probabilistic thinking. Presenting multiple perspectives* (pp. 417–439). New York: Springer.

Raoult, J. P. (2013). La statistique dans l'enseignement secondaire en France [Statistics in secondary teaching in France]. *Statistique et Enseignement, 4*(1), 55–69. http://publications-sfds.fr/ojs/index.php/StatEns/article/view/138.

Renyi, A. (1992). *Calcul des probabilités* [Probability calculus]. Paris: Jacques Gabay (Original work published 1966).

Savage, R. (1994). The paradox of nontransitive dice. *American Mathematical Monthly, 101*(5), 429–436.

Senior Secondary Board of South Australia (SSBSA). (2002). *Mathematical studies curriculum statement*. Adelaide, Australia: SSBSA.

SEP. (2011). *Plan de Estudios, Educación Básica*. México: Secretaría de Educación Pública.

Shaughnessy, J. M. (1977). Misconceptions of probability: An experiment with a small-group, activity-based, model building approach to introductory probability at the college level. *Educational Studies in Mathematics, 8*, 285–316.

Shaughnessy, J. M. (1981). Misconceptions of probability: From systematic errors to systematic experiments and decisions. In A. Schulte (Ed.), *Teaching statistics and probability: Yearbook of the National Council of Teachers of Mathematics* (pp. 90–100). Reston, VA: National Council of Teachers of Mathematics.

Stohl, H. (2005). Probability in teacher education and development. In G. Jones (Ed.), *Exploring probability in schools: Challenges for teaching and learning* (pp. 345–366). New York: Springer.

Stohl, H., & Tarr, J. E. (2002). Developing notions of inference with probability simulation tools. *Journal of Mathematical Behavior, 21*(3), 319–337.

von Mises, R. (1952). *Probability, statistics and truth*. London: William Hodge. (Original work published in 1928).

Wassong, T., & Biehler, R. (2010). A model for teacher knowledge as a basis for online courses for professional development of statistics teachers. In C. Reading (Ed.), *Proceedings of the 8th International Conference on Teaching Statistics*. Lubjana, Slovenia: International Association for Statistical Education.

Wild, C., & Pfannkuch, M. (1999). Statistical thinking in empirical enquiry. *International Statistical Review, 3*, 223–266.

Further reading

Batanero, C. (2013b). Teaching and learning probability. In S. Lerman (Ed.), *Encyclopedia of mathematics education* (pp. 491–496). Heidelberg, Germany: Springer.

Borovcnik, M., & Peard, R. (1996). Probability. In A. Bishop, M.A. Clements, C. Keitel, J. Kilpatrick, & C. Laborde (Eds.), *International handbook of mathematics education* (pp. 239–288). Dordrecht: The Netherlands: Kluwer.

Chernoff, E. J., & Sriraman, B. (Eds.) (2014), *Probabilistic thinking. Presenting multiple perspectives*. New York: Springer.

Kapadia, R., & Borovcnik M. (Eds.) (1991). *Chance encounters. Mathematics Education Library vol 12*. Dordrecht: The Netherlands: Kluwer.

Jones, G. A. (2005). *Exploring probability in schools. Challenges for teaching and learning. Mathematics Education Library vol 40*. New York: Springer.

Jones, G., Langrall, C., & Mooney, E. (2007). Research in probability: responding to classroom realities. In F. Lester (Ed.), *Second handbook of research on mathematics teaching and learning*. Greenwich, CT: Information Age Publishing and NCTM.

Jones, G. A., & Thornton, C. A. (2005). An overview of research into the learning and teaching of probability. In G. A. Jones (Ed.), *Exploring probability in school: Challenges for teaching and learning* (pp. 65–92). New York: Springer.

Shaughnessy, J. M. (1992). Research in probability and statistics: Reflections and directions. In D. A. Grouws (Ed.), *Handbook of research on mathematics teaching and learning* (pp. 465–494). New York: Macmillan.

Shaughnessy, J. M. (2007). Research on statistics learning and reasoning. In F. K. Lester (Ed.), *Second handbook of research on mathematics teaching and learning* (pp. 957–1009). Charlotte, NC: Information Age Publishing.

Author note: We appreciate the input from Juan D. Godino who gave critical feedback on this manuscript that greatly improved the final version. Financial support was provided by the Spanish Ministry of Economy and Competitivity (Project EDU2013-41141-P).